HOT VEGETABLE GARDENING:

Growing Early Veggies Using Traditional Hot Bed Gardening Techniques

BY

JAMES PARIS

BLOG: WWW.PLANTERSPOST.COM

Published By

www.deanburnpublications.com

Contents

First Edition December 2014

ISBN-13: 978-1505514353
ISBN-10: 1505514355

Copyright Notice

Other Relevant Books By James Paris

Raised Bed Gardening 5 Book Bundle

Companion Planting

Growing Berries

Square Foot Gardening

Compost 101

Vegetable Gardening Basics

Small Garden Ideas

Straw Bale Gardening

Root Cellar Construction

James Paris is an **Amazon Best Selling Author**, you can see the full range of books on his Amazon author page at..

http://amazon.com/author/jamesparis

Introduction:

Getting access to fresh vegetables in the depth of winter, particularly in the cold Northern regions, can be a real challenge.

The good news however is that with the right application of gardening techniques for colder climates such as Hot Bed gardening practices, this challenge has an effective solution already worked out for you!

Introduced by the Romans to the UK over 2,000 years ago to satisfy their need for fresh vegetables in the cold British winters, the technique has been adopted by many countries throughout Europe and the United States.

By the late 18th century, the Parisians – who were masters of this technique by now - were importing huge quantities of fresh vegetables to the United Kingdom.

This of course caught the attention of the UK gardeners and they were soon producing early vegetable crops from their own 'French Gardens.'

But what about now? What about the future for Hot Bed Gardening practices?

Hot Beds – Past & Present

The Decline of Hot Beds:
With the advent of the superstores and refrigerated transport however, along with other considerations; this method became more or less redundant – at least amongst commercial growers.

The growth of mechanized transport in the early part of the 20th[th] century, meant that horse manure was not so readily accessible. This being the main staple of the Hot Bed meant in turn that it was not commercially viable anymore.

New nitrogen-rich chemical fertilizers and modern ways to grow on commercial scales at reduced costs, without the need for all the preparation time involved with HBG, further reduced the need for Hot Beds of the organic type (see later explanation).

These 'advances' along with the simple fact that fewer and fewer individuals grew their own vegetables, choosing instead to lift them from the supermarket shelf at a lower cost most times than they could grow them!

The family unit changed dramatically after the two world wars. More women were now going out to work, rather than stay at home "tied to the kitchen sink " And throughout most of the Western world both men and women single or married, harboured ambitions to purchase their own house.

More money for living, less time for gardening. The bottom line is that it became cheaper and easier for consumers to simply purchase vegetables rather than get involved with actually growing them.

But that's not the end of the story!

The Advance of Hot Beds:
Now however we are seeing a massive return to the gardening practices, that had been more or less abandoned except by the enthusiastic few. Small gardening ideas in particular, such as Raised Bed Gardening, or Container Gardening, have come back into fashion as health conscious and environmentally aware individuals seek 'new' ways to grow their own vegetables.

The reasons for this positive shift in attitude towards growing vegetables I believe, comes from a combination of factors including but certainly not exclusively the following..

1: Environmental concerns. Many people are now very aware of the damage done to the environment by the use of chemical fertilizers and pesticides. Not to mention the 'carbon footprint' left behind owing to the transportation methods used over sometimes vast distances.

2: Health concerns. The chemical pesticides and fertilizers employed in commercial farming, do not only effect the soil, but varying quantities of these chemicals leech into the veggies themselves.

This leads to consumers perhaps consuming more than they bargained for, when eating commercially grown food. This of course includes animal products as well as vegetable.

3: Economic concerns. Although it is sometimes quite expensive to start a vegetable garden from 'scratch' (including tools & equipment etc), the actual cost of growing vegetables from seed is many times less than purchasing from the shops.

Add to this the fact that you know exactly what the growing conditions have been for your home-grown vegetables, and you can come to the conclusion that it is very cost-effective to grow your own – especially if you bring health concerns from eating contaminated products into the equation!

Conclusion:
Organic gardening methods in particular have grown in importance as all these, and many other considerations are taken into account by consumers.

Folks with limited time on their hands, and budgets as well as family health issues to consider, are looking for effective techniques to grow good healthy vegetables at reasonable cost and minimal effort.

Enter the no-dig gardening methods such as Cold-Frame Gardening, Raised Bed Gardening; Straw Bale Gardening; and Square Foot Gardening – all of which can incorporate one or more elements of **Hot Bed Gardening**.

As I hope to highlight in the next chapters – this method of growing vegetables in cold temperatures is definitely making a come-back!

What Is A Hot Bed Garden?

Also referred to as 'forced gardening' In a 'nutshell' a Hot Bed Garden is simply a growing area that has the soil heated by natural or artificial means, to make it possible to grow plants that would otherwise not grow owing to cold conditions.

As they say however 'the devil is in the detail' and so a longer explanation is perhaps required at this stage :)

Traditional 'natural' Hot Beds are beds heated by piling up 18-24 inches of manure or other organic material, to encourage decomposition. This is topped by 4-6 inches of good soil or growing compost.

The process of decomposition causes the organic material to heat up, which in turn warms the growing medium to a point where plants will germinate and thrive.

Other methods that do not require manure or organic material, involve running either electric heating cables under the soil to increase the temperature, or the use of hot water pipes or even open fires to produce warm air heating underground.

All of these methods will be discussed in greater detail in later chapters on the subject.

Hot Beds are used to lengthen the growing season significantly by making it possible to start very early, and finish very late in the season.

 In the UK for instance it is possible to be growing fresh veg from early-January to late November by using a Hot Bed.

In fact the only real restriction regarding growing times is the lack of daylight hours. As soon as at least 6-8 hours of natural daylight is assured however, then let the planting begin!

As briefly mentioned earlier, the big advantage in using Hot Bed gardening techniques however, comes when they are combined with the most popular no-dig ideas such as the ones mentioned earlier.

This is 'forced gardening' at its best, and the results can be quite spectacular and long-lasting – long after the Hot Bed cools down.

There are in fact many ways to construct your Hot Bed garden. The following chapter on how to create your own, will hopefully give you some ideas on constructing something to suit your own requirements.

NOTES/TO-DO PAGE

Creating A Hot Bed:

Natural Hot Beds

The traditional Organic Hot Bed begins with a Cold-Frame (sometimes known as a Warm-Frame). This is because the simple design and concept of the cold-frame, actually suits a Hot Bed Garden ideally.

Traditionally cold-frames are designed to keep in heat from the sun in the early part of the growing season, when the heat is weakest and the daylight hours shortest.

When utilising this idea with the Hot Bed concept however, we are relying on the Hot Bed itself to produce the heat that will become trapped inside the confines of the frame.

The picture below gives a clearer idea of what this means in real terms.

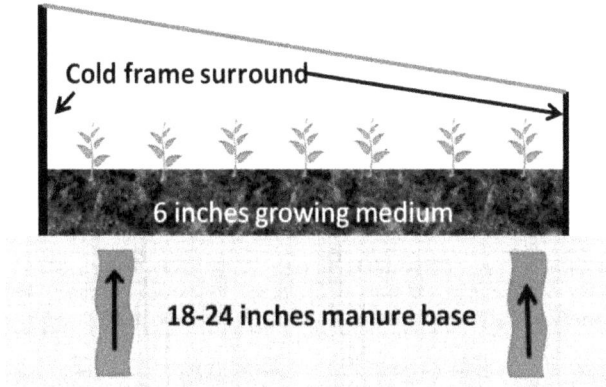

Cold frame surround

6 inches growing medium

18-24 inches manure base

In this picture we can see that the cold-frame is sitting on a base of horse manure. This is what will generate the heat which in turn will warm the soil and heat up the inside of the cold-frame, creating the ideal growing conditions for your vegetables – even in the coldest weather.

No matter how we lay out, or construct our Hot Bed garden, the idea of organic material producing enough heat to 'do the business' is essential for the whole operation.

The cold-frame itself is a simple structure consisting of a wooden frame slightly higher at the back and facing South. This can be topped with a glass-filled window frame, or a sheet of polycarbonate twinwall – which is far lighter and more effective for insulation.

The frame for a Hot Bed is best to be a movable simple structure, rather than fixed in place. This will allow for easy access to your Hot Bed material when it needs replenishing.

Typical measurements for a cold-frame are 4x3 or 3x3, any larger than this makes them more unmanageable and the plants difficult to reach.

The picture above is of a row of frames with one structure but multiple openings, as this following picture shows clearly.

This arrangement can only be achieved with a fixed permanent structure owing to the sheer size of it. However it is extremely effective, and is used on a semi-commercial basis to produce an excellent variety of vegetables throughout the year.

For normal domestic needs however, a simple wooden frame with a lid that drops down on top will do the job just fine.

Temperature:
The inside temperature of the cold-frame should be kept around 60F for spring vegetables. This can be achieved by simply opening the lids or 'lights' on sunny days to regulate or lower the temperature if need be.

The ground temperature inside will of course be largely dependent on the effectiveness of your Hot Bed arrangement, but this also is controlled by opening or closing the lights.

Step 1 – Collecting Organic Material:

To begin with, a good source of organic material is essential. The top choice for this is fresh horse manure mixed with straw from the stable. Horse dung is rich in nitrogen and the straw provides the carbon element. This makes Horse manure a 'hot' manure and the top choice for you're Hot Bed.

Second choice would be perhaps sheep, poultry (very hot), goat, or cattle.

I would advise against using pig manure or indeed any pet dung at all. Dung from carnivores can be full of parasites and other harmful larvae that are not present in 'veggie manures.'

Third choice should all of the above be out-with your ability to source, would be to use a mixture of any green organic material in the form of grass trimmings, vegetable waste etc, mixed with carbon materials such as paper (not glossy), cardboard, fine twig cuttings or dry leaves.

If you can lay hands on some nitrogen rich material such as poultry manure, great! Just mix it through the pile.

This material should all be piled up and a process of hot composting begun (more details on hot composting later), before laying out the bed itself.

Step 2 – Producing The Heat:

Once you have sourced your organic material, then you must prepare it for use in the Hot Bed. The first thing to bear in mind is that in the early stages of decomposition the material gets <u>very</u> hot, in fact it can reach over 150F quite easily.

If you were to use this super-hot material, then you would simply kill the plants – not exactly what you're hoping to achieve :)

There are two ways to do this. The first way is to take your manure from a fresh pile that has been sitting for a week to ten days.

GREAT STUFF – A STEAMING HOT PILE OF MANURE!

This is when the hottest period should be over, and the material inside the pile is beginning to cool down to manageable levels of around 80-90F or so.

Alternatively you can get a supply of fresh manure and pile it up nearby with a hose-down to start the process. After 7-10 days it should be ready to use.

If however horse manure or farm manure of any kind is not available to you, there is another method that will get almost as good results and that is Hot Composting.

Hot Composting is something that is in fact the subject of another book entirely (yea, you guessed it – I have one available!), however for this sake of this work I will include the basics of it here.

Normally used to speed up the composting process (you can produce usable compost in 3-4 weeks), it is nevertheless applicable in this instance as it also produces huge amounts of heat as well as nitrogen rich compost for vegetable growers.

Here is an overview of how to produce hot compost from my book Compost 101. Bear in mind that we will be using the material before it has finished the composting process described here.

Preparation:

First construct your material in a bin or open pile, no less than 3 x 3 x 3 feet. This is the minimum volume requirement for hot composting to take place. If it is an open pile then a covering such as a plastic sheet will keep in the heat and encourage the process.

Materials should be layered <u>as in the picture below</u> with a mixture of **nitrogen bearing materials** (grass clippings, fruit & vegetable scraps, herbivorous animal manure etc), and **carbon materials** such as sawdust, cardboard, dried leaves, wood chippings, twigs or straw.

Regarding layering of material – all compost is better layered or tiered, to introduce air and encourage the composting process – with perhaps the exception of layer composting or mulching which can include a single layer in certain circumstances.

In the following picture that shows the typical layout of a good compost heap, you will notice a layer of soil? This is not necessary in a hot composter, and indeed can make it heavier and hence more difficult to turn over. However the addition of manure is welcomed overall.

It is also worth noting that In the *Berkeley Method* the percentage of carbon bearing materials to nitrogen, should be around 30:1 for best results. If there is too much nitrogen bearing materials then the compost will decompose very rapidly leading to a loss of valuable nitrogen into the air.

All the materials should be chopped or chipped down into small pieces of about 1 ½ inches for best (and quick) results.

Top Layer - Leaves

Wet Greens

Dry Leaves

Water Well

Manure or Soil

Wet Greens

Dry Leaves

Water Well

Manure or Soil

Wet Greens

Dry Leaves

Branches or Pallet

After the material has been assembled into something resembling the picture above (not to scale!) then it must be left for 4 days without disturbance. This will allow the process to begin and the heat to build up inside the pile.

After this time, turn the compost so that the outside is inside and the inside out – I'm sure you get the drift!

The idea is that the material that has been on the outside of the pile and there-fore not exposed to the heat, is now in the center of the pile.

This process of turning should be done every day for the next 2 weeks. If turning every second day then a further week should be added to the time.

Turning the compost in this way prevents the heat build-up from becoming counter-productive for the composting process, as well as aerating the pile.

Using a 'compost bay' system **such as the one above,** makes turning the compost easier and more efficient as you flip it from one bay into another.

However this bay system is highly effective for both hot and cold composting methods.

Using this hot composting method should result in usable compost within 18-28 days – far shorter than the traditional cold compost pile method.

For our Hot Bed purposes, this system is best followed for about the first 7-10 days only. This ensures that the process is still 'active' and means that heat will still be produced when it is laid in the ground as the decomposition process continues.

It should be mentioned however that this hot compost will not last nearly as long as a Hot Bed made from manure, which can last from 2-3 months producing heat.

Hot Bed systems:
When it comes to the 'container' that holds this Hot Bed system, there are four main systems in use – and many variations of them.

There is of course a fifth option, which is an uncovered bed suitable for forcing on late spring vegetables without the need for an insulating cover.

However the four covered systems that I will go over in this book are Cold-Frame, Raised Bed, Square Foot, and Straw Bale gardening, which in itself is a type of container.

NOTES/TO-DO PAGE

Hot Bed Systems:

Cold-Frame System.

Whilst the organic material is doing its thing, the actual Hot Bed area can be made ready. With the traditional cold-frame system, this is done by digging out the area to a depth of 18-24 inches (450-600mm). By whatever size your frame will be plus 6 inches all around – as per the diagram above.

If you have poor drainage then dig out a sump area in the center (as in the picture below) about 1 foot deep, and fill with crushed gravel to catch the excess water.

Alternatively – and this is especially if you already have the cold-frame fixed in place – just dig out the inside area of the frame to within a 2-3 inches of the inside walls, to prevent possible collapse of the sides

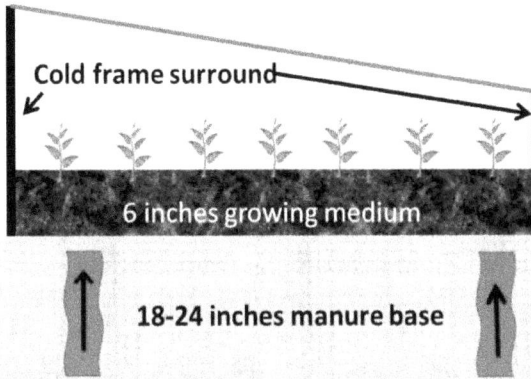

Cold frame surround

6 inches growing medium

18-24 inches manure base

When this is complete, then fill in the trench with your manure or compost mix. Press it down firmly all over to compress the material. This will remove some of the air trapped inside and in effect slow down the decomposition processes.

If the manure is excessively dry then give it a good soaking at this stage to encourage decomposition and heat build-up.

Cover the entire area with 4-6 inches of good quality growing medium. The mix used for a Raised Bed System such as 1/3rd compost 1/3rd peat & 1/3rd vermiculite or pearlite would be ideal.

This mix is light to work with, will not compact easily as will soil in a closed area such as this, and especially with the manure base, has all the nutrition your plants will need.

In both these examples it is also recommended to pile up soil or straw around the outside of the frame-work, or even builders insulating board. This will help insulate the frame and prevent heat loss during the colder weather.

Once this process is completed you are ready to begin your planting. (More on this in later chapters).

Raised Bed System.

Another popular way to use a Hot Bed is in conjunction with a typical Raised Bed system. This is simply a wooden frame laid on the ground with typical dimensions being 6 foot by 3 foot by 6-24 inches deep.

Usually with a Raised Bed, the area is completely filled with a good quality growing medium with a mix similar to the one mentioned earlier. In this instance however there is a change inasmuch as 2/3rds of the mix will actually be your manure.

TYPICAL COMMERCIAL RAISED BED SYSTEM

As you can see from the picture of a larger more commercial bed, this example is 2 foot high. This allows for 18 inches of manure and a 6 inch topping of composting mix. In the early cold months this has to be covered to retain the heat from the Hot Bed system.

This system is very versatile and can easily be adapted to cater for different weather or insect conditions over the whole of the growing season.

In this example below, a Raised Bed that has been covered in polythene in the early months to form a polytunnel above the Hot Bed, has now been covered with insect mesh as the season begins to warm up and the Hot Bed becomes less effective as the decomposition process slows down..

This loop system is very simply made with plumbers 1 inch plastic pipe, slipped down into a larger diameter pipe fixed to the inside of the frame-work.

The polythene or indeed insect mesh can be stretched out and fixed to a length of 3 x 2 and secured along the top of the base in such a manner that means it can be lifted for access.

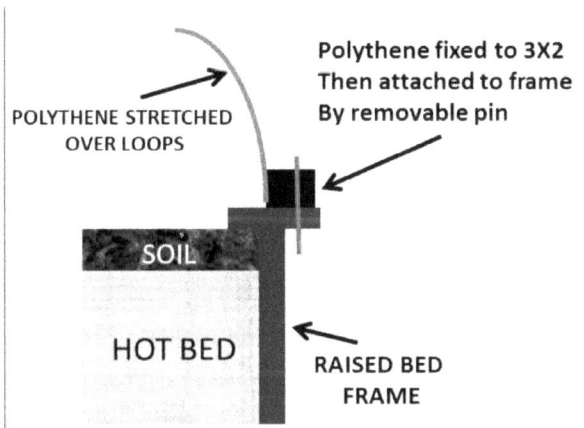

POLYTHENE STRETCHED OVER LOOPS

Polythene fixed to 3X2 Then attached to frame By removable pin

SOIL

HOT BED

RAISED BED FRAME

This same system can be used with a surface mounted Raised Bed system as in this example, or the frame-work can be sunk to whatever depth you want and utilised similar to the cold-frame concept.

To make a quick stackable Raised Bed system, that can be moved around after the growing season, or indeed raised to whatever height you desire (up to 2 foot max) – simply copy the following diagram making 6 inch high sections at a time.

This is a simple system using 6 inch x 1 inch timbers, fixed to 3 x 3 corner posts.
These sections are fitted one on top of the other and secured in place by the corner posts as per the diagram below.

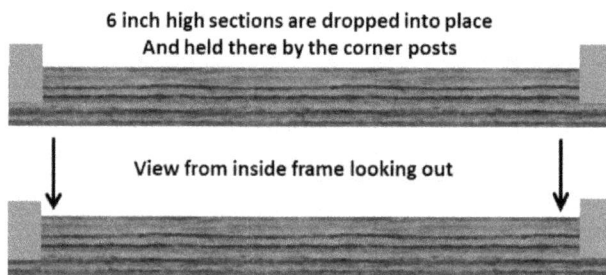

6 inch high sections are dropped into place
And held there by the corner posts

View from inside frame looking out

The complete frame is built 6 inches high at a time, and simply dropped one on top of the other until the desired height is reached.

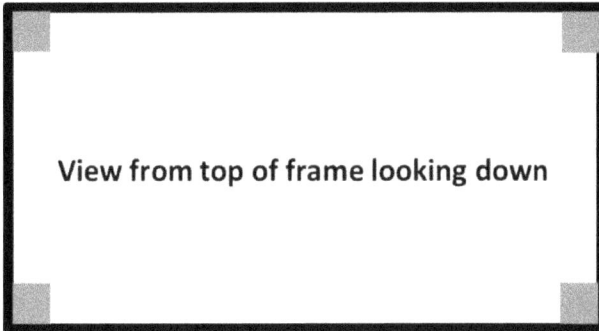

View from top of frame looking down

The essential finishing part in all this is that the growing area is covered to allow the inside space to reach the desired growing temperature for the vegetables planted.

Square Foot System:

Many readers will have come across the Square Foot Garden system, and if not…I have a book on this also!

However in a nutshell, the SFG is similar to the Raised Bed system in that it involves growing in a Raised Bed situation.

However the SFG is 4 x 4 which measures in at 16 square feet. Each of these 1 foot square areas is marked out and a variety of plants grown in the spaces allocated.

This system is more involved with super-effective plant rotation and growing medium, that maximizes the potential of this tiny area. This means that it is able to supply a small family with all their vegetable needs, over the growing season.

This is also an excellent system for companion planting techniques to help control pests and promote nutrient sharing between different varieties of vegetables.

The diagram below shows a typical SFG layout.

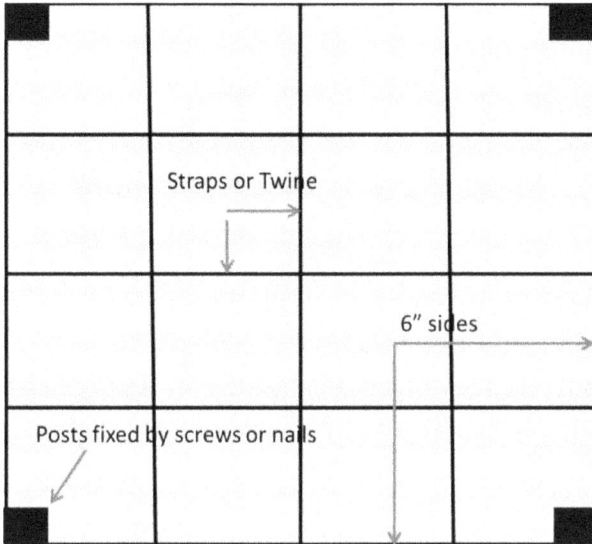

Similar to the conventional Raised Bed system, this can be built in accordance to the previous diagram with the only real difference being the dimensions of 4 x 4 instead of 6 x 3 which is the typical Raised Bed.

One Final Point regarding the 'surface' system and the 'sunken' system of frame-work. Once a Bed gets to 2 foot high, it is in danger of becoming unstable, or bulging at the sides due to the pressures exerted.

However a frame that is at least partially sunken into the ground has 2 main advantages over the frame that is sitting atop the ground.

1 – A sunken frame is much more insulated against the cold, especially a penetrating cold wind.

2 – The sunken frame is much more stable owing to the nature of the construction and surrounding groundwork.

With all Raised Bed systems I would also advise that you line the inside with a polythene membrane. This will both prevent the timber from soaking up the moisture and therefore drying out quickly.

It will also prevent any contaminants from treated timber (if indeed you are using such), from leaking into your growing area.

Straw Bale Hot Bed

Growing vegetables in straw bales is perhaps not immediately recognised as a type of Hot Bed Gardening. However this particular concept definitely falls into this category as will be explained..

First of all we take a typical straw bale measuring around 3 foot long by 18 inches high and 14 inches wide.

Straw Bale Gardening is a concept that is quickly gathering a following amongst vegetable gardeners in particular, as it is easy to operate. There is very little soil or compost to purchase – making it very economical. And it is very productive in respect to crop yield – if properly prepared!

To Begin..

First of all the bale must be 'primed.' This process is necessary for two reasons. To begin the whole decomposition process, and to fill the bale with enough nutrients to feed the growing plants.

Priming the bale is done by adding a nitrogen-rich fertilizer then soaking it thoroughly. The following instructions for priming and setting up a Straw Bale Garden are from by book on the subject.

<u>EXCERPT FROM STRAW BALE GARDENING BOOK – BALE PREPARATION</u>

Setting up/Feeding:

When placing the bale itself, it is important that it is arranged cut side up, or on edge. This enables the water to seep right to the inside of the bale and begin the 'cooking' process.

The 2 strings that hold the bale in place should be running along the sides. It is important that you leave these strings in place – for obvious reasons!

This 'cooking' is where the bale begins the process of decomposing and thereby producing heat in the inside of the bale, which in turn breaks down the straw and prepares it for your plants to benefit.

Once you have the bale in place then you must feed and water it to begin the process. Feeding is very important as the straw bale itself is just carbon (hay differs as per the previous chapters) , and your vegetables will need a good mixture of nitrogen and potash to develop fully.

This feeding can be achieved in several ways. Either through conventional fertilisers or store-bought organic fertiliser or (my preferred method), home-brewed organic feeds.

The **store bought nitrogen-rich feeds** should be spread over the bale at the beginning and watered in. Thereafter added to the water itself is often best before the bale is soaked.

Organic home-brew is simple to make and apply. Preparation should be done a few days before you are ready for the bale.

Add a few good handfuls of grass, stinging nettles, or borage which has been cut into pieces, to a pail of water. Weigh it down with a heavy stone or brick, then leave for 7 days to infuse.

The resulting liquid can be diluted at roughly 1 part feed to 10 parts water, then applied to the bale. The remaining liquid can be topped-up over the season with more plant material and water, and applied as necessary.

Compost feed can be produced by taking a few handfuls of compost and adding to water and left to infuse as previously described.

Manure feed is especially rich in nitrogen which your veggies will love, however it is the 'stinky' option! To make this tea place a shovel of manure (horse, sheep, rabbit chicken, or goat is ideal) into a hessian sack and put inside a deep pail of water for 5-7 days.

Squish the sack up and down a bit before removing from the pail (you can add it to the compost heap). The resulting tea should be watered down about 15 parts water to 1 part tea.

Be aware however that using fresh manure does carry an element of risk with regard to E.coli and other harmful bacteria and worm larvae that may be present.

With that in mind do take precautions when handling fresh manure, or alternatively (and safer) use well composted manure that has been composting for at least 1 year.

Fish meal, bone meal and seaweed also are great sources of nitrogen, potassium and other nutrients that will benefit your vegetables.

Priming The Bale:

The actual process of priming your bale should be done in the following sequence..

Day 1: Soak the bale completely with water infused with nutrients, or prepare the bale by scattering some store-bough fertiliser over it before soaking.

Day's 2-5: Continue with the soaking and feeding process. Monitor the internal temperature with a thermometer (a compost or meat temperature probe is ideal), and watch for the rise in temperature as the 'cooking' process begins.

Day's 6-14: Water every alternate day, checking to see that the bale does not dry out. As the process of cooking out comes to an end, the bale will cool down to reflect a temperature just a little higher or equivalent to, the ambient external temperature.

This means that the bale is ready to plant. If the reading is still high then wait till it drops before attempting to plant, otherwise it will be too hot for the roots and the plant will likely suffer a premature death!

**

Straw Bale Planting:
The main difference when using this method of HBG compared to the traditional, is that the planting really begins after the bale has significantly cooled down, otherwise the plants would simply burn.

Also you cannot rely on the heat of the bale to counteract the cold weather, as the heating process does not last long at all, compared to a manure Hot Bed.

That said – you can easily convert your straw bale into a mini-polytunnel by copying the following diagram of a straw bale covered to get the maximum benefit of the heating process.

If this is started at the right time of the season you will be able to plant earlier to take advantage of the heat in the bale, whilst capitalising on the warming sunlight as the season progresses.

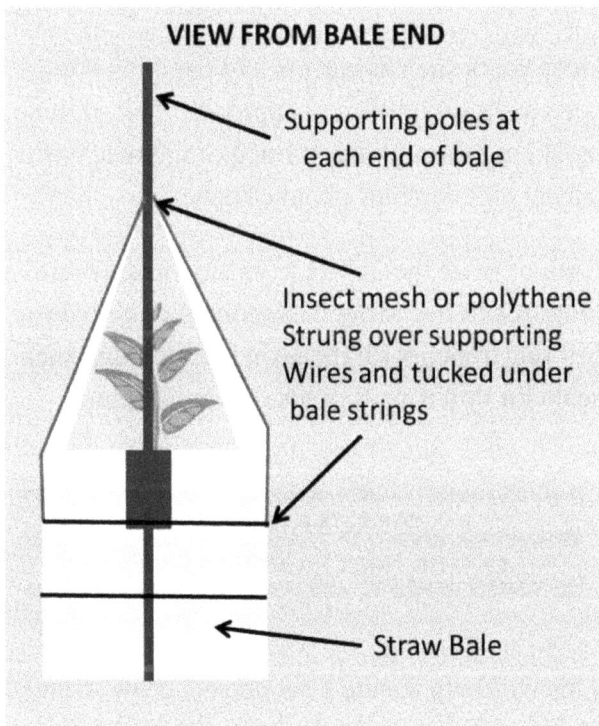

VIEW FROM BALE END

Supporting poles at each end of bale

Insect mesh or polythene Strung over supporting Wires and tucked under bale strings

Straw Bale

Planting in a straw bale is done by digging a hole a few inches deep and 4 inches or so around, then filling with a compost mixture to set your seedling into.

Alternatively you can plant from seed by covering the bale with two inches of soil/compost, and planting as you would normally.

Deep roots such as carrots and parsnips though need a slightly different approach. This is done by slicing through the surface of the bale with a trowel to a depth of about 6 inches.

Push or twist the trowel from side to side until a long trench has formed as in the picture below. Fill this trench with the growing medium, then scatter a thin row of seeds down the center.

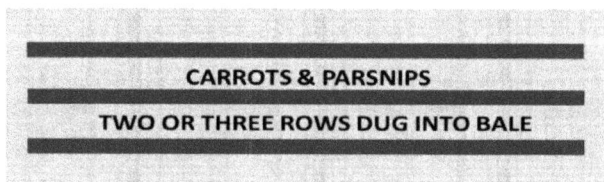

CARROTS & PARSNIPS

TWO OR THREE ROWS DUG INTO BALE

This will help ensure your carrots grow straight and strong down into the bale without growing all weird as in the next picture!

Although not fully a 'Hot Bed' as in the earlier descriptions of such; this can be an extremely effective 'type' of Hot Bed gardening which if you have never tried it, I would highly recommend for the following reasons.

1: It is perhaps the most authentic of the no-dig methods as it requires virtually no digging at all.

2: It is extremely productive and versatile regarding the vegetables that can be grown throughout the season.

3: Like all forms of Raised Beds, there is no problem with back-ache as bending over is at a minimum!

4: Cheaper to use than traditional Raised Bed gardening as little or no compost is used, and enclosing in a frame is optional – thereby saving on timber. Straw bales can vary in price between $2-$7 each or , but they are certainly not expensive as such.

5: It's a great talking point with friends and family, most of whom will never have heard of the idea – let alone seen a working example!

6: easier to keep slugs and snails at bay as they are not keen on the sharp straw – at least in the early stages of the growing season.

Other Hot Bed Systems

If you are looking for alternatives to the Organic Hot Bed system then there are other options available for you to consider, including electrical, hot water and hot air systems.

The advantages of these ideas varies according to which one you choose, but basically they offer the possibility of better heat regulation than the organic method.

They also take away the need for a supply of fresh horse, or other manure to get started. Of course you will need other supplies of different material according to your choices – life is a bit of a trade-off sometimes!

Electrical Hot Bed

Perhaps the most regulatory of the systems. This idea basically means that you are totally in control of your Hot Bed temperature at all times. Most cables come with fitted thermostats set at about 70F.

The downside of course is that you need to have access to an electricity supply. The system may also be quite expensive to purchase and to operate over time with an average cable cost of around $1.50 - $2.50 per linear foot.

That said, it is very easy to fit and operate. To do this simply follow the directions above but without the need for an 18 inch deep manure pit.

Once you have purchased your electrical heating, then lay it out on the surface in accordance with the manufacturer's instructions. Take care that the cable does not cross over itself particularly, as it is likely to melt!

Cover over with a thin 1 inch layer of sand then weed fabric, before covering with 4-6 inches of soil or growing medium. The weed fabric will prevent you from accidentally pulling up the cable whilst working your veggie patch!

Fluid Hot Bed:

This can be constructed exactly the same way as the electrical bed, but using hot water pipes instead. This is usually the least preferred or indeed applicable choice as it depends on a ready supply of hot (or at least warm) water.

TYPICAL HOT WATER SETUP

Hot water can be supplied by means of a ground source heating pump, if you are able to supply this, or indeed from any other water heating system such as your home central heating or solid-fuel furnace.

Another method is to supply the hot water using solar panels. However the shortage of sufficient sunlight during the shorter winter days to heat the water, makes this idea mostly redundant.

Soil temperature can be controlled with the use of a thermostat sensor installed into the soil, and able to shut off the supply. This water system is best operated using narrow, insulated central heating pipe.

An extremely effective solution, it is nevertheless dependent on your ability to supply constant hot water to your growing area.

Hot Air System.

This is perhaps along with the manure bed, one of the oldest Hot Bed systems still in use today. It was a particular favourite of the Romans and was extensively used by the Victorians in their heated greenhouses and particularly their heated fruit walls.

The theory behind the idea was simple. Create an air space underneath the growing beds or through walls, through which warm air could pass and heat the soil above it.

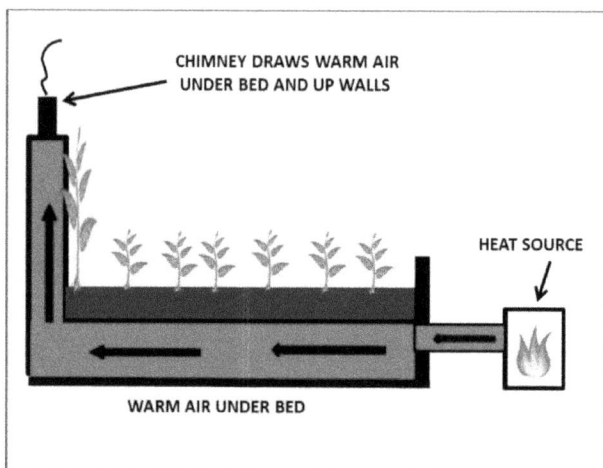

CHIMNEY DRAWS WARM AIR UNDER BED AND UP WALLS

HEAT SOURCE

WARM AIR UNDER BED

Indeed the same system was used in Roman times to heat the houses and bathing areas of the wealthy – an early example of under-floor heating!

Old as this system is however, it is largely redundant owing to the more manageable and cheaper systems now at our disposal.

Summary:
No matter what system you choose, all of these systems have a common requirement – the need to insulate against heat loss. This will both improve your growing efforts, and reduce overall costs of heating.

When using underground cables or water pipe, you can reduce this loss by laying down builders reflective insulating board before applying the heating pipes/cables.

Insulating your plants against severe night-time temperature drops is also essential. This can be done by covering the plants over with garden fleece or hessian sacking at night-time.

You can also cover the cold-frame with a sheet of insulating board over-night to keep out the cold – remember to remove it at first light though!

A Hot Bed can also be used to heat a greenhouse or polytunnel quite effectively – at least to keep the interior frost-free in the main.

In fact, running a Raised Hot Bed along even one side of the Polytunnel in this way, can help you produce crops throughout most of the year.

Planting Out Your Hot-Bed

When planning out the plants for your Hot Bed, 3 main things have to be taken into consideration.

1: The number of daylight hours – This has to be 6-8 hours minimum. This is not so important if you are using artificial lighting to supplement the daylight hours..

2: The time each plant takes to reach maturity, as well as the temperatures in which they will thrive.

3. The depth and hence lifespan of the heat produced through your Hot Bed.

The daylight hours requirement is I think self-explanatory, so no need to labour the point here. I would just add that if you want to grow out-with the required daylight hours, then the use of artificial lighting such as fluorescent lights are a "cheap and cheerful" option.

With regard to the growing times required for the individual plants to reach maturity, the following chart should help out in this regard.

VEGETABLE	MINIMUM TEMP. F	OPTIMUM TEMP. F	APPROX. DAYS TO HARVEST
HARDY VEGETABLES			
BROCCOLI	40	80	65
CABBAGE	40	80	85
KOHLRABI	40	80	50
ONIONS (set)	35	80	65
LETTUCE (leaf)	35	70	60
LEEKS	40	80	120
PEAS	40	70	65
RADISH	40	80	30
SPINICH	40	70	40
TURNIP/SWEDE	40	80	50
SEMI-HARDY			
BEETS	40	80	60
CARROTS	40	80	70
CAULIFLOWER	40	80	65
PARSLEY	40	75	80
PARSNIPS	35	70	70
POTATOES	45	80	125
SWISS CHARD	40	85	60

With the Hot Bed manure set at a depth of about 18 inches, this means that heat will be produced for 2-3 months. This results in most vegetable planting starting in mid to late January in the UK and other cold Northern regions throughout the USA and Europe.

Planting out a Hot Bed is usually done in 4 stages, really to match the seasons. Though this will of course vary according to individual needs and climatic circumstances.

First Stage:

Typically this would begin in January by planting salad seedlings that have been forced on in a greenhouse over December. Carrot seeds for instance can then be sown between them. This allows for the maturing of the salads before the carrots are ready to lift.

Salad such as lettuce (leaf varieties), golden frill, rocket, fennel and of course radishes, are ideal for an early start to the season.

First early potato varieties such as Pentland Javelin, Orla, Arran Pilot and Rocket, can all be planted in a hot Bed to get an even earlier potato harvest.

On a point of note here. If you are sowing carrots amongst leaf vegetables such as lettuce etc, then to avoid disturbing the soil around the carrots snip the other vegetables away with sharp snips when ready to harvest, rather than lifting them out.

Second Stage:

After around 6-8 weeks, as the soil in the Hot Bed begins to cool you should be approaching the slightly 'warmer' weather of early spring. This is the ideal time to plant out the cool season crops such as broccoli, cabbage, swedes, parsnips and cauliflower.

These are planted early so that they are ready for harvest before the weather heats up, as this would encourage them to 'bolt' and become inedible or extremely stringy and tough at best.

Second early potatoes such as Jazzy, Charlotte, or Maris Peer should be planted around this time.

Frost is still very much a threat this early in the year, so be sure to protect your young plants in the evening by covering over with a protective fleece or sacking material.

Third Stage:

By late Spring the bed is cooled down but is full of nutrients such as nitrogen and potash from the partially decomposed manure below the soil.

If you are using a Cold-Frame with a removable lid, then this is the time to consider lifting it away to allow the growth of the young tomato plants, peppers, courgettes, cucumber, pumpkin, leeks and onions – all of which will benefit from the nitrogen-rich manure base of the bed.

Main crop potatoes such as Desiree, Setanta, Maris Piper, and King Edward can also be planted at this time.

Early Autumn gives the opportunity to plant more cool season varieties such as winter cabbage, broccoli, brussels sprouts, and winter harvesting leek varieties.

<u>Fourth Stage:</u>

By the end of the year the bed should be empty and young seedling being forced-on in a heated greenhouse, ready to be planted in your new Hot Bed!

Dig out the old material which is now reduced to rich mainly composted material – ideal for digging into the traditional veggie patch or Raise Bed garden.

Now you are ready to begin the whole process all over again – wasn't that a quick year :)

Authors Note

Finally I would just like to say a HUGE THANKS for purchasing my book, and I truly hope that you have found it to be beneficial and perhaps a little inspiring :)

When it comes to planting vegetables, or indeed any plant, so much is dependent on local conditions both with regard to weather and soil conditions. There is indeed no "1 size fits all" when it gets down to it.

However provided all the ingredients are in place regarding water, sunlight, temperature, and nutrients; then with a little application your plants will thrive – no "green fingers" required!

Now for the shameless plug :)

I have an extensive collection of gardening related books available through Amazon, especially relating to ideas for small vegetable gardening practices.

A few of these are listed below, however you can see the full range on my Amazon authors page.

Other Relevant Books By James Paris

Raised Bed Gardening 5 Book Bundle

Companion Planting

Growing Berries

Square Foot Gardening

Compost 101

Vegetable Gardening Basics

Small Garden Ideas

Straw Bale Gardening

Root Cellar Construction

James Paris is an **Amazon Best Selling Author**, you can see the full range of books on his Amazon author page at..

http://amazon.com/author/jamesparis

NOTES/TO-DO PAGE